REPORTS FROM
THE BOTANICAL INSTITUTE, UNIVERSITY OF AARHUS

NO. 15

Current Scandinavian Botanical Research

in ECUADOR

Edited by

Benjamin Øllgaard & Ulf Molau

1986

Botanisk Institut
Aarhus Universitet
ISSN 0105 - 4236
ISBN 87-87600-19-6

CONTENTS

3.4.1.2. Jan-Erik Bohlin, GB: Taxonomic Investigations in the Nyctaginaceae in NW South America.

3.4.1.3. Uno Eliasson, GB: The Amaranthaceae in the New World; morphological characters and taxonomic relations.

3.4.1.4. Bente Eriksen, AAU: Pollen and Fruit Morphology in the Genus Valeriana

3.4.1.5. Jens E. Madsen, AAU: Taxonomic Problems in Ecuadorean Cactaceae.

3.4.1.6. Ulf Molau, GB: A Monograph of the Genus Bartsia.

3.4.1.7. Bertil Ståhl, GB: Theophrastaceae and the Sexual Systems in Clavija.

3.4.2. Monocotyledonous Angiosperms

3.4.2.1. Anders Barfod, AAU: Phytelephantoid Palms - Taxonomy and Ethnobotany

3.4.2.2. Roger Eriksson, GB: Systematic and Morphological Studies on Cyclanthaceae, especially the Genus Sphaeradenia.

3.4.2.3. Mats Hagberg, GB: A Revision of the Genus Monotagma (Maranthaceae).

3.4.2.4. Lauritz B. Holm-Nielsen, AAU: Distribution of the Alismatidae in the Neotropics.

3.4.2.5. Flemming Skov, AAU: Revision of the Palm Genus Hyospathe in Tropical South America.

3.5 Vegetation Studies:

3.5.1. Jørgen Korning, AAU: New Results in the Anango Rain Forest Project

3.5.2. Simon Lægaard, AAU: Páramo Vegetation in Ecuador

3.5.3. Karsten Thomsen, AAU: Preliminary Results of Floristic Studies of Rain Forest Trees in Anango.

3.6. Applied aspects

3.6.1. John Brandbyge, AAU: Reforestation by means of Local Tree Species in Rural Districts of Ecuador

3.6.2. Lars Peter Kvist, AAU: Aspects of Ecuadorean Lowland Indian Ethnobotany.

3.6.3. Bo Boysen Larsen, AAU: Prospects of Quinua Cultivation in the Andes.

4. Summary of Recent Activities

4.1. L.B. Holm-Nielsen, AAU: The Aarhus University Ecuador Project (AAU Ecuador Project) 1984-86, with Maps of Collection Localities.

4.2. Ulf Molau, GB: The Gothenburg University Neotropical Program 1984-86, with Maps of Collection Localities.

4.3. B. Øllgaard & K. Molau: A Cumulative Map of the Collection Localities of Scandinavian Botanists in Ecuador.

5. Lauritz B. Holm-Nielsen: Concluding Remarks.

6. List of Staff and Students involved in the Ecuador Projects

1. Introduction

For many years the botanical institutes of the University of Aarhus (Denmark) and Göteborg (Sweden) have maintained a fruitful collaboration in the study of Ecuadorian flora and vegetation. In the spring of 1984 participants of the Ecuador projects of the two institutions met in Central Jutland, Denmark, in order to present papers on their current research and exchange ideas. The proceedings of that workshop were published as an issue called "Scandinavian Botanical Research in Ecuador" edited by L.B. Holm-Nielsen, B. Øllgaard and U. Molau (Reports Bot. Inst. Univ. Aarhus, No. 9. 1984). Due to the success of the first workshop it was decided to arrange similar meetings every second year.

The second meeting was held in the rough idyll of Bohus-Malmön in the parish of Bohuslän in the skerries of the Swedish west coast from 24 to 26 of April 1986 with 27 participants. The present issue is a report from that workshop, containing abstracts of the papers presented and reports on the status of various projects and on collection activities during the last two years.

We are deeply indebted to "Per Adolf Larssons Stipendiefond" for financial support to the meeting. We highly appreciate the excellent organisation by Jan-Eric Bohlin who provided a very pleasant setting for the meeting.

We wish to express our gratitude and indebtedness to our many collaborators in Ecuador, whose active support was essential for the results of our work in the country. We are especially grateful for help from the Departamento de Biologia, P. Universidad Católica del Ecuador, Quito, The Departamento de Parques Nacionales y Vida Silvestre (MAG), The Museo Antropologico del Banco Central, Guayaquil, PREDESUR, Loja, and the Departamento de Botanica, Fisiología y Sistematica, Universidad Nacional de Loja.

2. Gunnar Harling, GB:

Flora of Ecuador - Its Present Status

The preliminary work for a Flora of Ecuador began in 1968 and the first part (Cyclanthaceae) was published in 1973. Since that time 23 parts of the Flora have been published, the latest of which is Polypodiaceae-Asplenioideae by Robert G. Stolze. In proofs or ready for printing are the following families: Eremolepidaceae, Viscaceae and Loranthaceae by Job Kuijt; Begoniaceae by Lyman B. Smith & Dieter C. Wasshausen; Alismataceae, Limnocharitaceae, Hydrocharitaceae, Juncaginaceae, Potamogetonaceae, Zanichelliaceae, and Najadaceae by Lauritz B. Holm-Nielsen & Robert R. Haynes; Dicksoniaceae, Lophosoriaceae, and Cyatheaceae by Rolla Tryon; Pontederiaceae by Charles N. Horn; Amaranthaceae by Uno Eliasson; Anacardiaceae by Anders Barfod; and Passifloraceae by Lauritz B. Holm-Nielsen, Peter Møller Jørgensen & Jonas Lawesson.

The number of published families and families ready for publication is just now 50 out of 230. A further 65 families are under active work and many of them will appear during the next few years. It is, however, somewhat embarrassing that we have up to now not found collaborators for several large families, e.g. Caesalpiniaceae, Papilionaceae, Euophorbiaceae, and Piperaceae.

The number of vascular plant genera in the Ecuadorean flora can be estimated to about 1,800-1,900. of these 330 have been published or are ready for publication. The total number of species is very difficult to estimate for the moment, but may be as high as about 20,000. So far 2,160 species (and a large number of subspecies and varieties) have been published or are ready for publication in the

Flora.

I have also tried to estimate how many collection numbers
the floristic work is based on and have come to an
approximation of about 250,000 for the time being. About
half of these collections are located in the herbaria of
AAU, GB and S. Comparatively may be mentioned that the flora
of Peru was originally based on about 33,000 sheets, but
this figure has of course been multiplied since that time.

A further comparison with the Flora of Peru may be of
interest. The publication of this Flora started in 1936 and
has, with an interruption between 1971 and 1980 continued
ever since. The Flora is now near completion, although
several large families, e.g. Acanthaceae, Gesneriaceae,
Asclepiadaceae, and most parts of the Compositae, are still
lacking. All the families are under active work, however, so
it may be anticipated that the Flora of Peru will be
completed before the end of this century. According to
Gentry (Flora of Peru, Fieldiana Bot., N.S., No. 5, 1980,
pp. 1-5) 11,789 species of spermatophytes have been treated
to date and slightly more than 2,000 additional species will
appear in the still unpublished families. Moreover, about
800 species of pteridophytes will be added. Since the
earlier parts of the Flora are very incomplete owing to the
scarcity of collections, Gentry´s guess that the Peruvian
flora in reality consists of "well over 20,000 species"
seems rather probable. Paying attention to the fact that
Peru is more than four times as large as Ecuador one may get
an understanding of the extreme wealth of the Ecuadorian
flora since, as mentioned above, also this is expected to
contain about 20,000 species of vascular plants.

Colombia, the other neighbouring country of Ecuador, has
undoubtedly a much greater number of plant species than Peru
and Ecuador. Owing to the fact that a modern Colombian flora
is still lacking, an estimate of the species number must be
hihgly uncertain, but guesses of about 50,000 species have
appeared in the litearture. If this is true, Colombia would
have about the same number of species as the whole of
Brazil.

3.1.1. Thomas Læssøe, AAU:

Mycological Studies in Ecuador

From early June to early September 1985 I spent my second period of field work in Ecuador. The purpose was mainly to collect xylariaceous fungi for my graduate work in mycology. A total of 519 collections were gathered, including a few lichens and vascular plants (e.g. Thismiaceae and Balanophoraceae). Duplicates of pyrenomycete samples were mailed directly to Liverpool Polytechnics, where Dr. A.J.S. Whalley brought some material into culture, being most successful with Xylaria and Kretzschmaria and less so with Hypoxylon. Continued study of herbarium specimens plus the new samples give a total of approximately 40 Xylaria species from Ecuador (10 thus far unidentified).

I have tried to get other fungal groups determined by various experts. The following groups have been or will be submitted full or in part to the named workers. So far no publications dealing with this material have appeared:

Xylariaceae: A.J.S. Whalley (Liverpool); J.D. Rogers (Washington State University).

Hyprocreaceae s.l.: C.T. Rogerson (New York Botanical Garden).

Cordyceps: R.A. Samson (Centraalbureau voor Schimmelcutures, Baarn); H.C. Evans (Commonwealth Inst. of Biological Control).

Corticiaceae s.l.: K. Hauerslev (ass. Botanical Museum, Copenhagen).

Poroid Aphyllophorales: L. Ryvarden (Botanical Institute, University of Oslo).

Ramaria: J.H. Petersen (Botanical Institute, University of

Aarhus).

Agaricales: D.N.Pegler (Kew); C. Bas (Rijksherbarium, Leiden); S.A. Elborne (Botanical Museum, Copenhagen); E. Rald (ass. do.).

Heterobasidiomycetes p.p.: F. Oberwinkler (Lehrstuhl spezielle Botanik, University of Tübingen).

Myxomycetes: H.F. Gøtzsche (Institut for Sporeplanter, University of Copenhagen).

The remaining groups have not (or virtually not) been studied. Besides a paper on Xylaria, papers on Hypoxylon, Kretzschmaria, Thamnomyces and Camillea are planned, p.p. in collaboration with A.J.S. Whalley. There are preliminary plans for a flora project and for an illustrated guide to the more common and characteristic macrofungi.

Elisabeth Bravo of Universidad Catolica, Quito, is for the moment studying mycology with J. Hedger (University of Aberrystwyth) on a British Council grant. Hopefully this will lead to an increased local interest in mycology.

3.2.1. Lars Arvidsson, GB:

The Lichen Flora of Ecuador

Ecuador has a richly diverse and well-developed lichen flora which, however, has not until recently been investigated in any depth. A literature guide (Hawksworth 1977) records only two works dealing with this subject, viz. Müller Argoviensis 1879 and Zahlbruckner 1905. Both contributions contain notes on small lichen collections by other botanists. In addition, a check-list of lichens from the Galapagos was published by Weber in 1966. Very occasionally lichens from Ecuador are cited in modern monographs and revisions.

In 1972 Mr. Dan Nilsson and I undertook an excursion to Ecuador, one of the first specialized in lichens. However, there exist also earlier scattered collections (herb. S) made by Swedish phanerogamists, viz. E. Asplund and G. Harling. A survey of recent expeditions of lichenological interest by botanists from Göteborg is shown in tab. 1. In the last two decades the lichen flora of Ecuador has received an increased interest from various collectors and important material is held also in e.g. AAU, BM, COLO, NY and U.

Our collecting work excludes the Galapagos and was concentrated to the Andes (fig. 1), particularly to the humid forests between 1,000 and 4,000 m altitude. Many genera of great size and beauty are met with here, viz. Sticta, Lobaria, Pseudocyphellaria, Leptogium, Pannaria, Erioderma, Coccocarpia, Hypotrachyna, Everniastrum and Usnea. This luxurious flora demanding constant oceanic conditions is very sensitive to air pollution and deforestation and therefore worthy of special attention. The eastern lowlands (El Oriente) and the coastal plain (La Costa) are poorly known lichenologically. However, some isolated visits there indicate an impoverished lichen flora.

Even though many crustaceous species were gathered, the activity has primarily been focused on foliose and fruticose taxa. Epiphytic species are more frequently collected than terricolous and saxicolous ones. The subsequent identification work was assisted by many specialists. Thus, difficult genera like Cladina, Cladonia, Parmelia, (s.lat.), Collema, Stereocaulon and others are already named and in many other groups determination work is going on. Crustaceous lichens constitute a great problem as for many genera there is no competence available. Even though frequently collected, groups like e.g. the pyrenolichens remain unidentified. A survey of the genera so far detected in Ecuador is given in tab. 2 together with notes on collaborators, number of specimens and species. The herbarium in GB comprises about 10,000 specimens of lichens. Over 130 genera are hitherto known from Ecuador and this number is very likely to swell as the work proceeds. An estimation of the total number of species in the country is very difficult to make at this stage. However, what can be said with certainty is that the lichen flora does not at all contain such a multiplicity of taxa as for instance the phanerogams. In spite of the astonishing diversity of habitats that exists within this country the number of lichens is probably similar to that of Scandinavia, that is about 2,000 species, or even less than that. A recent flora of New Zealand (Galloway, 1985) also a country with a wide range of environments, records 966 taxa in 210 genera. That figure is supposed to represent about 60% of the lichens to be found there.

The GB material constitutes an excellent base for a lichen flora of Ecuador. The aim is to publish on various groups as soon as they have been determined. Such contributions shall be illustrated and follow a standard format. The lay-out is not appointed in detail as this, at least to some extent depends on the place of publication. A great problem is that there is no permanent lichenologist at the institute in Göteborg. Much of the work has to be done during week-ends and the future of the project appears somewhat uncertain.

Lichens are much more sensitive to changes in their environments than many other plant groups. Construction of new roads, cutting of forests and atmospheric pollution will put many Ecuadorian lichen communities at risk, or have already done so. This development is very likely to increase and extend into the areas still untouched by man. Mapping of the lichen flora of Ecuador is therefore a very urgent task.

COLLECTOR	YEAR	NO. OF NUMBERS
L. Arvidsson & D. Nilsson	1972	1400
L. Andersson	1974	350
L. Andersson, U. Molau & M. Neuendorf	1977	200
L. Arvidsson & D. Nilsson	1979	900
I. Andersson	1980	200
M. Lindström	1982	150
L. Arvidsson & A. Arvidsson	1983	2200
L. Arvidsson, M. Lindström & M. Lindqvist	1985	2000
M. Lindström, L. Arvidsson & M. Lindqvist (only Leptogium)	1985	600
M. Lindqvist, L. Arvidsson & M. Lindström (only Sticta)	1985	500
L. Andersson	1985	300
	Total:	8800

Tab. 1. Survey of the lichen collections from Ecuador in GB. In addition there are also scattered collections by other botanists and several duplicates from AAU. Some 300 numbers can be added. The number of specimens is about 10.000.

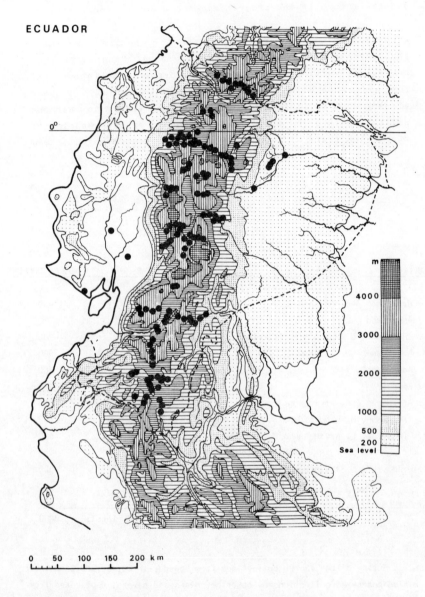

Fig. 1. Localities visited by swedish lichenologists during the period 1972-1985.

THE LICHEN FLORA OF ECUADOR (MATERIAL IN GB)

GENUS	CONTRIBUTOR	NUMBER OF SPECIMENS	EST. NUMB: OF SPECIES	MATERIAL DETERM. AND RETURNED
Acarospora		1	1	
Alectoria		11	1	
Anaptychia	R. Moberg, Uppsala	150	?	
Anzia	I. Yoshimura, Kochı	25	1-2	
Anthracothecium		2	1	
Arthonia		7	?	
Arthrorhaphis		5	2	
Aspicilia		1	1	
Asterothelium ?		1	1	
Aulaxinea		1	1	
Bacidia		17	?	
Baeomyces	H. Sipman, Berlin	60	5	*
Biatorella		1	1	
Bombyliospora	J. Poelt, Graz	5	?	
Brigantiaea	J. Hafellner, Graz	10	1	*
Bryoria ?				
Buellia	K. Kalb, Obermarkt	5	?	
Bulbothrix	M. Hale, Washington	1	1	*
Byssolecania		1	1	
Byssoloma		1	1	
Calenia		1	1	
Calicium	L. Tibell, Uppsala	4	?	
Caloplaca		54	?	
Candelaria	J. Poelt, Graz	5	?	
Candelariella	-"-	4	?	
Candelina ?	-"-			
Catillaria	H. Sipman, Berlin	8	?	
Catinaria	-"-	10	2	*
Catolechia		1	1	
Cetraria ?	I. Kärnefelt, Lund			
Chiodecton	G. Thor, Stockholm	30	?	
Chroodiscus		1	1	
Chrysothrix ?		7	?	
Cladia	T. Ahti, Helsingfors	47	2	*
Cladina	-"-	55	5	*
Cladonia	-"-	100	19	*
Coccocarpia	L. Arvidsson, Göteborg	270	7	*
Coelocaulon ?	I. Kärnefelt, Lund			
Coenogonium		10	2-3	
Collema	G. Degelius, Göteborg	30	5	*
Cryptothecia		1	1	
Dactylina		1	1	
Dictyonema		65	3	
Dimerella		13	2-3	
Diploschistes	T. Lumbsch, Marburg	22	2-3	
(Dirina)				
Dirinaria	A. Aptroot, Utrecht	10	?	
Echinoplaca		1	1	
Endocarpon		4	?	
Erioderma	L. Arvidsson, Göteborg P.M.Jørgensen, Bergen	300	13	
Everniastrum	H. Sipman, Berlin	150	9	*
Everniopsis	M. Hale, Washington	3	1	*

Flavopunctelia	M. Hale, Washington	20	1	*
Glossodium		2	1	
Graphina ?		25	?	
Graphis ?		30	?	
Gyalectidium	R. Santesson, Uppsala	2	1	
Haematomma		5	1	
Heterodermia	R. Moberg, Uppsala	150	?	
Hypogymnia		1	1	
Hypotrachyna	M. Hale, Washington	260	32	*
Leioderma	P.M.Jørgensen, Bergen	1	1	*
Lecanora		15	?	
Lecidea		12	?	
Lepraria		7	2-3	
Leprocaulon		15	2	
Leproplaca		2	1	
Leptogium				
sect. Lept. etc.	M. Lindström, Göteborg	800	20	
sect. Mallotium	P.M.Jørgensen, Bergen	200	9	.
(Leptotrema)				
Lobaria		200	6-8	
Lopadium		2	1	
Mazosia	R. Santesson, Uppsala	1	1	
Megalospora	H. Sipman, Berlin	25	3	*
Menegazzia	P. James, London	5	2	
Micarea ?		1	1	
Nephroma	P. James, London	4	?	
Neuropogon	J. Walker, London	12	2	*
Normandina		25	1	
Ochrolechia		4	?	
Oropogon		80	2-3	
Pachyospora		5	1	
Pannaria	P.M.Jørgensen, Bergen	100	7-8	
Parmelia	M. Hale, Washington	4	1	*
Parmeliella	P.M.Jørgensen, Bergen	100	5	
Parmelina	M. Hale, Washington	15	7	*
Parmotrema	-"-	140	26	*
Peltigera	O. Vitikainen, Helsingfors	110	10	
Peltula ?	B. Büdel, Marburg	1	1	
Pertusaria		14	?	
Phaeographis ?		20	?	
Phaeophyscia	R. Moberg, Uppsala	10	?	
Phlyctis		1	1	
Phlyctella?		1	1	
Phyllophiale	R. Santesson	1	1	*
Physcia	R. Moberg, Uppsala	30	?	
Physcidia ?		1	1	
Physciopsis	R. Moberg, Uppsala	5	1	
Physconia ?	-"-	5	?	
Phyllopsora	L. Brako, New York	8	?	
Placopsis		17	3-5	
Placynthium		1	1	
Polychidium	L. Arvidsson, Göteborg	8	1	*
Porina		3	?	
Pseudocyphellaria	D. Galloway, London	150	7	*
Pseudoparmelia	M. Hale, Washington	7	5	
Psora	E. Timdal, Oslo	2	1	
Psorella	L. Brako, New York	2	1	

(Psoroglaena)				
Psoroma		1	1	
Punctelia	M. Hale, Washington	35	5	*
(Pyrenula)				
Pyxine	A. Aptroot, Utrecht	15	?	
Ramalina		90	10	
Rhizocarpon		1	1	
Rinodina		7	?	
Roccella		16	1	
Sphaerophorus	L. Tibell, Uppsala	11	1	
Siphula	R. Santesson, Uppsala	3	2	
Solorina		2	2	
Sporopodium		1	1	
Stereocaulon	H. Sipman, Berlin	140	16	*
Sticta	M. Lindqvist	1000	30	
Strigula		5	?	
Tapellaria		1	1	
Teloschistes	O. Almborn, Lund	80	3	*
Temnospora		7	1	
Thamnolia	R. Santesson, Uppsala	13	2	
Thelotrema	M. Hale, Washington	3	?	
Toninia		1	1	
Tricharia		1	1	
Trichothelium		1	1	
Umbilicaria		5	?	
Usnea		300	25	
Xanthoparmelia	M. Hale, Washington	6	5	*
Xanthoria		26	2-3	
Xylographa		1	1	

Tab. 2. List of lichen genera so far known from Ecuador. This survey is based on the material in GB. The collections made during 1985 (about 3500) are not included. A few genera recorded by Weber (1966) from the Galapagos, but not with certainty detected in the mainland are put in parenthesis. 136 genera are presented here but the number will increase when many of the microlichens (e.g. pyrenocarpous taxa, foliicolous species etc.) have been investigated more carefully.

3.2.2. Lars Arvidsson, GB:

Revision of the lichen genus Erioderma

A revision of the lichen genus Erioderma is in preparation. Twelve years ago, Mr. Dan Nilsson at GB started this work but left it unfinished some years later. The present investigation is a joint project together with Prof. Per Magnus Jørgensen, Bergen, Norway, who for a long time has taken a great interest in this genus and related groups.

Species of Erioderma are foliose and lobate, medium-sized to rather large lichens attached to the substrate by rhizines. A distinctive feature of the genus is the hairy upper surface of the lobes. Some taxa have stiff, erect hairs, other have an arachnoid tomentum. The lower surface is either more or less naked (in some species with distinct, raised veins) or covered with rhizines. The thallus is heteromerous but lacks a lower cortex. All species have blue-green photobionts, viz. Scytonema. Apothecia are lecanorine and usually marginal. The asci contain 8 simple, colourless spores. Erioderma belongs to the Pannariaceae and is related to Leioderma, Pannaria and Parmeliella.

Erioderma has a mainly tropical montane distribution and is most frequent at altitudes between 1,000 and 3,000 m. It prefers humid habitats such as cloud-forests and is always epiphytic, often preferring rather narrow branches of low bushes. Only one species, viz. E. pedicellatum (E. boreale) reaches the boreal zone in the northern hemisphere.
Very little has been written about this genus. One reason is very likely the scarcity of material and it is often stated that if you are lucky to find a species of Erioderma it will only be a few scrappy thalli. However, during various excursions to the eastern cordillera in Ecuador we have come across rich communities with up to 6 or 7 sympatric species. Even though a pantropical genus it appears that Erioderma has a centre of variation in the northern parts of the Andes. At the moment we are distinguishing 25 species throughout the world, 14 of these undescribed. In Ecuador we

have so far found 14 species, 9 of which are new. The GB
herbarium contains about 300 specimens of Erioderma. This
remarkable collection (mostly from Ecuador) has been a
starting-point for the revision.

As the chemistry of Erioderma was completely unknown, we
started with a thin layer chromatography of all specimens.
We soon realized that this was a very complex field and many
unidentified spots appeared in the chromatograms. Dr. Jack
Elix, Canberra, Australia, recently came to our succour and
his more sophisticated methods like high performance liquid
chromatography, nuclear magnetic resonance and mass
spectroscopy gave very interesting results. About 20
different compounds were detected by him, of which a
majority is undescribed. For instance, there are several new
depsides and depsidones, a new chemosyndrome and chlorinated
barbatates, an unknown yellow pigment (isomeric with
vinetorin) etc. The biochemical relationships of these
products is also known in most cases and has of course been
taken into consideration. The distinction of species so far
is based on gross-morphology together with chemistry and
phytogeography. In many cases a distinct chemistry also
corresponds with a distinct morphology. However, in other
taxa (e.g. E. chilense) a uniform morphology has turned out
to be chemically complex. Here, the variation can be
correlated with differences in distribution indicating
"chemical taxa" at subspecies level. E. verruculosum
represents a third category in which it seems impossible to
couple a diverse chemistry together with distribution,
ecological preference or morphology. A yellow pigment in the
thallus of this species is seen in some lobes, but is absent
in other of the same thallus. In this species the chemistry
is of no taxonomic importance. The taxonomic interpretation
of the chemical variation in Erioderma is an intricate
problem which needs further research.

3.2.3. Mats Lindquist, GB:

The Lichen Genus Sticta in Neotropics

The lichen genus Sticta comprises conspicuous foliose species with dorsiventral, heteromerous thalli. It belongs to the family Lobariaceae (formerly Stictaceae), the closest relatives being Pseudocyphellaria and Lobaria.

Sticta species are usually rather large, with more or less broad lobes, and a polyphyllous or monophyllous thallus. The lower surface is distinctly pitted with holes or so-called cyphellae (a character that immediately reveals this genus). The photobiont is either green (Trebouxia/Myrmecia) or blue-green (Nostoc) algae.

Sticta is a cosmopolitan genus, mainly tropical, and houses about 200 species. It is favoured by humid habitats, thus being very common and important in the epiphytic vegetation in tropical montane forests, cloud and mountain rain forest in South America. In Ecuador, Colombia and Venezuela Sticta seems richly diversified.

An unusually large collection from Ecuador of this genus is kept at GB. The aim of my studies is to carry out a taxonomical revision of Sticta, or the Neotropical species. The genus has not been properly monographed since the 1900th century.

3.2.4. Marie Lindström, GB:

Taxonomical Studies in the Lichen Genus Leptogium

The genus Leptogium (Fam. Collemataceae) comprises about 150 species of mainly foliose, usually medium sized lichens with a homoiomerous, non-stratified thallus. The photobiont is the cyanobacteria Nostoc, the pectine-sheath of which makes the thallus gelatinous when wet. Leptogium is separated from Collema by the presence of a true distinct cortex of one, rarely two or three layers of irregularly isodiametric cells. Soredia do not occur within the genus, but a number of species have isidia. The ascospores of Leptogium are ellipsoid or fusiform, occasionally elongate-clavate in shape and have transverse and usually longitudinal septae. There are eight spores per ascus. The anatomy and ontogeny of the ascocarps provide important diagnostic characters. Other important specific characters are: shape and branching, surface texture, anatomy and thickness of the lobes; presence or absence of isidia; and shape, size, septation and ornamentation of the ascospores. Chemical characters are lacking.

Leptogium has a world-wide distribution, but reaches its greatest abundance in the tropical and subtropical parts of the world. It can be found in many different types of habitats, but since the photobiont is dependent on liquid water for its photosynthetic activity, it is most frequent in very humid habitats. In the Neotropics Leptogium is highly diversified in the montane rain forest, the cloud forest, and the lower and middle páramo. A few species are confined to the lowland rain forest. The altitudinal distribution ranges from sea level up to c. 4,000 m. Species of Leptogium are primarily corticolous, but saxicolous, terricolous, and rarely foliicolous species occur as well.

Leptogium is usually divided into five sections, of which Leptogium and Mallotium are the two largest. The latter differs from the other sections by the presence of small hairs, mainly on the lower side of the thallus. Among the tropical non-hairy species, two main species groups can be discerned, the L. azureum group and the L. phyllocarpum group. The first group is characterized by a more or less smooth thallus with distinct lobes, whereas the thallus of the second group always have longitudinally wrinkles and irregularly anastomosing lobes.

Leptogium has two geographical centers of variation, one in NE South America (The Andes) and one in SE Asia. Presently, I am working on a revision of the Neotropical species of the genus (excl. sect. Mallotium) including about 35 taxa. Evidence from herbarium material as well as field observations have been used. Field studies have been carried out in Ecuador, Venezuela, The Dominican Republic, and Florida, USA. The investigation has also included morphological analysis, anatomical and ultrastructural studies, as well as statistical calculations of spore measurements. Future plans are to use computerized programs for phenetic and cladistic treatments.

3.3.1. Benjamin Øllgaard, AAU:

Lycopodiaceae in Ecuador

With approximately 80 species Ecuador has about 1/4 of the total number species of the Lycopodiaceae. Together with Colombia, Ecuador is the center of diversity for the family. Three of the four genera which are now recognized in the family are represented in the area. Lycopodium (sensu medio) has 5 species belonging in 4 very distinct sections. There are problems with species distinction in the type section. Lycopodiella comprises 8 or 9 species in two sections. Lycopodiella alopecuroides is polymorphic, while distinction of the species in the group of L. cernua is based on characters of growth habit and size, mainly. A single species is exceptional, being epiphytic.

Huperzia is by far the largest and most problematic genus. No infrageneric subdivision is apparent in the species outside the group of H. selago. Morphologically and anatomically the genus is relatively undifferentiated, with very few specialized characters. Morphogenesis seems to be unstable in most species, and may be strongly affected by external factors. The descriptive tradition in the genus is very weak. Most characters are highly plastic within a species, e.g. stem thickness, number of leaf orthostichies, the gradual or abrupt heteroblastic leaf sequence, leaf direction, leaf margin characters etc. Hybridization seems to occur rather uninhibited, and contributes to the blurring of species limits. The putative hybrids often have normal meioses and normal spores. As a consequence, species recognition is often based on some experience and comparison with identified material, rather than a set of definite characters.

3.3.2. Suzanne Roth, GB:

The Fern family Gleicheniaceae in the Neotropics.

Gleicheniaceae is a pantropical family of primitive ferns, comprising about 120 species. In the New World they range from Mexico to S Chile, and are furthermore distributed in the Galapagos Islands, the Cocos Island, Juan Fernández, and the Falkland Islands. The aim of my studies has been to carry out a taxonomical revision of the Neotropical species of the family. In this area it is represented by two genera, Gleichenia Smith with about twentyfive species and Dicranopteris with five. Many of the species are pioneer ferns on exposed soil in the lowland rain forest and in the cloud forest where they can form huge hybrid swarms. A single species, Gleichenia simplex, occurs in the páramo vegetation. With regard to overall growth habit this species recalls the páramo species of other unrelated fern genera, such as Jamesonia and Polystichum, as obvious case of convergence.

Gleicheniaceae is a difficult group to revise taxonomically, especially since they are very hard to grow in the greenhouse. However, I have been able to propagate gametophytes from spores on soil in Petri dishes, and I have also found gametophytes of one species in the field.

3.4.1.1. Fanny Astholm, GB:

Genecological and taxonomical studies in Alonsoa (Scrophulariaceae)

Alonsoa is a small genus of about ten species distributed in the mountains of Central and South America. The highest species diversity within the genus is found in northern Peru. They are primarily found in disturbed areas like roadsides, scree areas, and on cultivated ground, between 1,000 and 3,500 m.s.m. The genus is included in the non-parasitic tribe Hemimeridae. The plants are most common as small shrubs or subshrubs, but perennial and annual herbs occur.

There are two species-complexes present in Ecuador, the widely distributed A. meridionalis group and the A. linearis group which is restricted to Ecuador and Peru.

The taxonomy of the genus is confusing. It is difficult to get hold of the morphological variation, and there is a need for a study based on the species-biological background. The aim is to solve the taxonomic problems with genecological methods combined with morphological studies.

Alonsoa is buzz pollinated like almost 8% of the Angiosperms. It is a unique method of pollen harvesting from apically dehiscent anthers. Bees release pollen by shivering their indirect flight muscles. The characteristic audible vibration (the reason for naming this buzz pollination) results in a rapid expulsion of the pollen. The bees carry a net positive charge and conditions are set for electrostatic attachment of the negatively charged pollen grains. This type of pollen collection is a fast and efficient method in an optimal foraging sense.

Within my study of <u>Alonsoa</u>, the following points will be of certain consideration:

- Comparative cultivation in greenhouse to study inter- and intraspecific morphological variation.

- Crossing experiments to study the relationships among different taxa.

- Investigation of the mating system, as an important background to the patterns of gene flow. Important subjects in this regard will be pollen/ovule ratios, pollen tube competition, incompatibility genes, and floral morphological differences between outcrossing and predominantly selfing populations.

- Optimal outcrossing distance. Seed set is measured after hand pollination with pollen from donor individuals at different distances from the female receptor parent. The vitality and fertility of the offspring is measured as well, as an estimate of the fitness of different genome combinations.

- Reward. Pollen is the only reward in buzz pollinated flowers. Consequently it is of interest to look at pollen chemistry and odour.

- Pollinator behavior, which strongly affects the overall pattern of gene dispersal in outcrossing populations. Studies of flight distances, directionality, handling time, specifity, and pollen carry-over are important parts of this project.

3.4.1.2. Jan-Eric Bohlin, GB:

Taxonomical investigations in the Nyctaginaceae in NW South America.

1. Colignonia

A monograph of the genus will soon be published. It comprises 6 species, 3 subspecies and 2 varieties. One species is described as new. In Ecuador 5 species are recorded. A great problem has been to find good morphological characters useful for a field-key. These polymorphic species often differ in microscopic details not useful in field. The two sections Colignonia and Pterocarpae are well defined by their number of perianth-lobes and their different chromosome numbers (5-lobed 2n=34 & 3-lobed 2n=32).

2. Neea

Species of the genus Neea are perennial, dioicous shrubs and trees, 1 to 40 m tall. They are restricted to the areas east of the Andes, with an altitudinal distribution from 1200 m and lower. They range from Venezuela and Colombia in the north, via Ecuador, Peru and Brazil to Bolivia in the south. The lateral dichasial inflorescences have small flowers, 3 to 4 mm long. The pistillate flowers are subglobose and the staminate flowers are more or less urceolate. The fruit is drupe-like with a more or less fleshy anthocarp. The leaves are opposite.

An expedition to the oriental parts of Ecuador and Peru is planned for the early 1987, with the main purpose to make special collections of the genus Neea.

Live material of this genus will be collected and is also requested from other parts of Neotropics.

3. Flora of Ecuador

The Nyctaginaceae of Ecuador comprises 10(11) genera and c. 30 species. The contribution to the Flora is under preparation and will be published as soon as the genus Neea has been revised.

3.4.1.3. Uno Eliasson, GB:

The Amaranthaceae in the New World; morphological characters and taxonomic relations.

Details of androecia and gynoecia are fundamental in the recognition of genera within the Amaranthaceae. It is suggested that androecia and gynoecia of different genera may be phylogenetically more closely related than would be suspected from a cursory examination. I hypothesize that the type of staminal tube found in Pseudogomphrena and Froelichia is derived from the type found in Alternanthera and Froelichiella by a reduction of filament length and a fusion of pseudostaminodia with the filaments. The staminal tube of Gomphrena could result from a further decrease in distance between pseudostaminodia of the Pseudogomphrena type, and a deeper forking of the pseudostaminodia. If this is the true evolutionary pathway, each so-called apical filament lobe in Gomphrena would be homologous with half a pseudostaminodium in Pseudogomphrena. In my opinion, the variation in the androecia among these and other genera, as well as within genera such as Pfaffia, can be interpreted as the combined results of coalescence and splitting-up tendencies.

The Amaranthus type of pollen is found in the majority of genera of the subfam. Amaranthoideae. A group of genera within the subfam. Gomphrenoideae also has pollen identical with or close to this type. Most genera of subfam. Gomphrenoideae have pollen of the Gomphrena type. The deviant and monotypic genus Pseudoplantago (Argentina) has uniloculed anthers (at anthesis), a characteristic of subfam. Gomphrenoideae, but floral morphology clearly connects the genus to a group of genera (among others

Achyranthes and Cyathula) within the Amaranthoideae. The
pollen grains of Pseudoplantago are cuboidal, a very unusual
shape among the angiosperms.

Floral morphology and palynological characters indicate a
close relationship between Alternanthera and Pfaffia. The
genera Woehleria (Cuba, monotypic) and Irenella (Ecuador,
monotypic) may be derived from, or be of the same origin as
Dicraurus and Iresine. All four genera are placed in the
subfam. Gomphrenoideae because of the bisporangiate anthers,
but the pollen structure is very close to the Amaranthus
type. A merger of the two last named genera might be
defended, but unless a complete generic revision of Iresine
would provide special support for a merger, I prefer to
recognize Dicraurus as a distinct genus, differing from
Iresine in the structure of the female flowers. Iresine has
been estimated to comprise more than 70 species, but the
genus is in bad need of revision, and the true number of
species is probably about 40. Many species are variable. It
is possible that different ploidy levels might explain the
morphological variation found in, for example, I. diffusa,
but very few chromosome counts have been made in the genus,
and some may be suspected to have been made on incorrectly
determined material. I. spiculigera has been accommodated
with I. diffusa due to a large number of intermediate
specimens. However, the variation in some morphological
characters in specimens of I. diffusa s. lat. does not
follow a normal distribution curve. This may indicate that
more than one taxon is involved or that there are different
ploidy levels within I. diffusa s. lat. In fact, a
chi-square goodness of fit test carried out on selected
floral characters convincingly indicates that the characters
do not follow the normal distribution curve expected in a
taxon with normal variation.

Twenty-three genera of Amaranthaceae occur in the New World.
Among the genera of the subfam. Amaranthoideae, Celosia,
Cyathula, and Achyranthes have their main distributions in
the Old World and are represented in the Americas only by
widespread weeds. All the New World genera of the subfam.
Gomphrenoideae are mainly or entirely restricted to this region.

3.4.1.4. Bente Eriksen, AAU:

Pollen and Fruit Morphology in the Genus Valeriana

Earlier classifications of the Valerianaceae have emphasized characters of fruit morphology and anatomy at the level of genus, and in some cases at the level of species. Also pollen characters show some generic differences, but earlier studies have failed to find any consistence in these structures, especially in the genus Valeriana. With new information of pollen and fruits it is my aim to divide the relatively restricted number of Ecuadorian Valeriana species into related groups, thereby also improving the sections proposed by Höck (1882) and Graebner (1906). This new information also leads to a better understanding of the species and the variation among them.

The treatment of the data is not yet finished, but the provisional results show some correlation. The pollen can be divided into 5-7 groups on basis of their ornamentation and size. Even though these groups do not fully correspond to the sections of Höck and Graebner, there is nothing really surprising in the results. The most puzzling element is that different species of twining shrubs have pollen belonging to different groups. The fruits are more difficult to categorize, but it is possible to distinguish some types. The shape, number of pappus rays and the haircovering, combined, essentially give the same sections as the pollen. In general, fruits carry more specific information.

3.4.1.5. Jens E. Madsen, AAU:

Taxonomic problems in Ecuadorian Cactaceae

The Cactaceae contain approx. 80 genera and maybe 2000 species. In Ecuador 13 genera and approx. 30 species are native.

A major problem of Cactaceae taxonomy is the poor represen- tation in herbaria. In order to provide better material it is recommended to collect living material of non-flowering plants, and to prepare herbarium vouchers from cultivated individuals. This is especially advisable for species with large, slender, nocturnal flowers.

The present distribution of the family in Ecuador is argued to reflect a floristically heterogenus origin. If the 12 Ecuadorian genera of the subfamily Cactoideae are treated in accordance with Buxbaum´s "Phylogenetic Division of the subfamily Cereioideae (e.g. Cactoideae)" (Madrono 14: 177-206. 1958) they must be placed into six out of eight possible tribes!

Ecuadorian terrestial cacti grow in several, isolated interandean deserts, at altitudes ranging from 700 to 3000 m. The most important deserts are from north to south: Chota Valley, Guayllabamba Valley, Chanchán Valley, Río León-Río Jubones Valley, Catamayo Valley. Borzicactus and Opuntia are the dominant cacti, to the south further Espostoa, Melocactus a.o. In the dry bunch grass páramo around Palmira Opuntia cylindrica is common up to 3500 m.

Terrestial cacti also grow at the coast in the desert and deciduous dry forest around the Santa Elena Peninsula, the Manta region and at Huaquillas. Armatocereus is very common, also in disturbed, semi-arid regions. Further Pilosocereus,

Monvillea, _Hylocereus_ and _Opuntia_.

Ecuador has a rich and diverse flora of epiphytic cacti, dominated by _Rhipsalis_, _Epiphyllum_, _Hylocereus_ and _Disocactus_. _Weberocereus (Eccremocactus)rosei_ is endemic. Epiphytic cacti grow in the lowlands along streams, especially in southern Ecuador, where they occur up to 2000 m. Several species are found in the Ecuadorian part of the Amazonian basin.

3.4.1.6. Ulf Molau, GB:

A monograph of the genus Bartsia

The genus Bartsia (Scrophulariaceae-Rhinanthoideae) is widespread in the alpine areas of the world. It totals about sixty species, the majority of which are found in the tropical Andes. In Africa the genus is represented by three species and in Europe by four species. In alpine Asia, Bartsia is absent, obviously replaced by Pedicularis, and the same case is seen in the Rocky Mountains. It appears to be a general rule that in areas where Bartsia is abundant, Pedicularis is rare, and vice versa.

Throughout its distribution, Bartsia is pollinated primarily by bumblebees (Bombus). In the tropical Andes, the genus Bombus is represented by a relatively now number of species, but instead there is a remarkable niche partitioning with regard to its relation to the Bartsia species. Thus, in Ecuador Bartsia laticrenata is pollinated by huge Bombus funebris queens, while the sympatric Bartsia melampyroides is pollinated by the much smaller Bombus funebris workers. In addition, some of the Neotropical Bartsia species are hummingbird pollinated.

In South America, the genus Bartsia is readily subdivided into four distinct sections, and this subdivision is supported by cytological evidence. With regard to phytogeography the overall pattern of Bartsia species distributions agrees well with that of other Andean Angiosperms with short-distance dispersal, such as Calceolaria, Alonsoa, Colignonia etc.

3.4.1.7. Bertil Ståhl, GB:

Theophrastaceae and the sexual systems in Clavija

The small family Theophrastaceae comprises five genera of Neotropical shrubs and small trees. The family is traditionally classified next to the Myrsinaceae and it is sometimes placed as a subfamily of it. The five genera are distinguished mainly by floral characters, and to some extent by habit. Deherainia (1 sp., S Mexico - Honduras) together with Jacquinia (c. 30 spp.) forms a group of "normally" branched shrubs and small trees. The remaining three genera, Neomezia (1 sp., Cuba), Theophrasta (2 spp., Hispaniola), and Clavija (c. 30 spp.), mostly consist of completely or almost unbranced shrubs or (sometimes in Calvija) small trees, with the leaves clustered at the top of the stems.

Jacquinia has its centre in the Antilles; 1 species (J. keyensis Mez) reaches Florida and several species are found in Central America as well as in the coastal regions of northern South America. The genus is restricted to dry, lowland areas, preferably close to the coast (cf. J. pubescens H.B.K. in Ecuador).

The genus Clavija is represented by one species on Hispaniola, three or four occur in Central America (Panamá - Costa Rica) and the rest are distributed over the South American mainland south to Paraguay and southern Brazil. In Ecuador 11 taxa have been recognized so far. The genus is found primarily in the lowland, inhabiting both wet and dry areas.

An interesting problem that indeed deserves further attention concerns the evolution of the sexual systems in Clavija. Whereas the other, mainly Antillean genera of the

family are hermaphroditic throughout, species of Clavija
exhibit various forms of dimorphic sexual systems.
Curiously, although perhaps purely incidental, these
evolutionary trends coincides with the geographical
distribution. The sole Antillean species, G. domingensis
Urb. et Ekm., has, as far as can be seen in herbarium
specimens, functionally bisexual flowers. By contrast, all
other species of Clavija, in Central America and in the
South American continent, have evolved two types of flowers,
female and male, fundamentally distinguished by free or
united stamen filaments, respectively. Although remaining
closed during the course of anthesis, the anthers of the
female flowers hold either pollen with high stainability (as
in the Ecuadorian C. repanda Ståhl) or pollen that stains
poorly and/or are irregularly shaped, thus indicating a
gradually reduced maleness, which, however, always result in
functionally female flowers. On the other hand, the ovary of
the male flower morph is sometimes entirely rudimentary and
empty (as in the Central American C. costariacana Pitt.),
but in most species the ovary of the male flower morph holds
several mostly fertile ovules. But then, of course, fruit
setting by the male flower morphs remains to be seen in
living plants. In this contect, the two species C.
longifolia (Jacq.) Mez and C. nutans (Vell.) Ståhl seem to
represent a rather extreme development. The two species are
among the most successful in the genus, C. longifolia being
present in Venezuela and are distributed east of the Andes
down south to Ecuador, whilst C. nutans is widely
distributed south of the Amazonas down to southern Brazil.
Despite an ample amount of herbarium specimens only one
collection of C. longifolia (from cultivation!) has been
found with flowers of the female morph. If the collections
reflect a real condition, both these species must have
evolved a secondary form of hermaphroditism, by means of a
flower morph, which in other species of the genus has become
purely male.

3.4.2.1. Anders Barfod, AAU:

Phytelephantoid Palms - Taxonomy and Ethnobotany.

Three genera make up the palm subfamily Phytelephantoideae. Phytelephas is the genus most rich in species. Fifteen species have been described, but a future revision of the subfamily will probably include much fewer species. Only one species, P. microcarpa R. & P. has thus far been collected in Ecuador. The genus Ammandra includes a single species, A. decasperma Cook distributed in the Chocó region. A new species was recently collected by Balslev and Henderson in southern Ecuador. The final genus Palandra includes the ecuadorian "Tagua"-palm, P. aequatorialis (Spruce)Cook. The hard endosperm of the seeds, well known as "vegetable ivory", was used for making buttons and handicrafts. Before the 2nd World War, Ecuador made a major part of their export earnings on this material. Later the vegetable ivory was replaced by cheap plastics.

A combination of primitive and advanced characters indicates a long evolutionary history of the Phytelephantoid palms. Ever since the first species of Phytelephas were described from Peru by Ruiz and Pavon in 1798, the position of this unique group has been debated. They have been referred to the closely related families Pandanaceae and Cyclanthaceae, or to a family of their own. Today they are considered true palms and the most recent classification gives them the rank of subfamily.

Like other palms, the Phytelephantoid palms are used for a great variety of purposes. The vegetable ivory was already mentioned. The fleshy tissue surrounding the hard endocarp and the fluid endosperm of the immature seeds are considered delicacies in many places in Ecuador. The newly emerged staminate inflorescences are sometimes dried and sold on markets as snacks. Also the palm-heart, or "palmito" is eaten. In areas where Phytelephantoid palms are common their leaves are used for thatching of roofs. The fibres extracted from the leafbases of some species are used for making brooms.

3.4.2.2. Roger Eriksson, GB:

Systematic and morphological studies in Cyclanthaceae, especially the genus Sphaeradenia

Cyclanthaceae, which has an exclusively Neotropical distribution, consists of perennial herbs, shrubs, lianas, and epiphytes. The family comprises about 200 described species in 11 genera. It forms a conspicuous part of the flora in humid vegetation types at low and medium high altitudes.

The inflorescence is an axillary peduncled spadix with pistillate and staminate flowers arranged in alternate cycles (subfam. Cyclanthoideae, usually considered monotypic) or in spirally arranged groups with one pistillate flower surrounded by four staminate (subfam. Carludovicoideae). The leaf blades are generally bifid (most genera), rarely flabelliform-parted (Carludovica) or entire (Ludovia, Pseudoludovia), with one to three costae. The genera in the "Asplundia group" (Carludovica, Asplundia, Thoracocarpus, Schultesiophytum, Evodianthus, Dicranopygium) have spirally arranged leaves and four parietal placentas and in the "Sphaeradenia group" (Ludovia, Pseudoludovia, Sphaeradenia, Stelestylis) distichous leaves and one to four apical to subapical placentas.

The genus Sphaeradenia (c. 50 spp.) is closely related to Stelestylis (4 spp.) and a group of taxa that probably forms a new genus, "Chorigyne" (c. 6 spp., 4 undescribed). Sphaeradenia and Stelestylis both have connate pistillate flowers, but the former has one apical placenta and seeds without appendages, while the latter has four apical placentas and seeds with appendages. "Chorigyne" also has seeds with appendages, but is distinguished from the other genera by having free pistillate flowers and four subapical placentas. Furthermore, the seed coat is differently developed in the different genera. Sphaeradenia has a western distribution from Costa Rica to Peru and Venezuela, with its centre of variation in Colombia. Stelestylis has an eastern range from Venezuela to Suriname, and "Chorigyne" a northern in Costa Rica and Panama.

The pollen morphology, studied with SEM, has given interesting results. The basic pollen type is monosulcate to monoulcerate, boatshaped and foveolate. But there is a variation in grain size, shape, aperture shape, aperture position, and exine sculpturing that has proved to be taxonomically useful at generic, or in the large genera, subgeneric levels.

3.4.2.3. Mats Hagberg, GB:

A Revision of Monotagma (Marantaceae)

Monotagma is a strictly Neotropical genus comprising 30-35 spp. with a distribution from Nicaragua in the north to Mato Grosso in the south. Monotagma has three centres of diversity: (1) The Napo-Iquitos area with 10 spp. of which 2 are endemic, (2) The Imerí area in southern Venezuela with 6 spp. of which 4 are endemic, and (3) The Manaus area with 6 spp. of which 2 are endemic. These areas are among those that have been proposed as Pleistocene forest refuges by various authors. Most of the species grow in the primary forest on low altitudes (to c. 1100 m). Some species are confined to the Amazonas caatinga/campina.

Monotagma differs from all other Neotropical Marantaceae genera by the one-flowerd cymules. All other genera have two-flowered cymules. The loss of a flower in a flower pair is a relatively "easy" step in microevolution, and one might suspect that this could have happened more than once. In fact there is an East Asian genus which also has one-flowered cymules, namely Monophrynium. But Monotagma is indeed homogeneous and forms a natural group. The shoot system is highly uniform: the rhizome is more or less trailing above ground supported by "stilt" roots; all species except sect. Ulei have a basal leaf rosette and a terminal synflorescence on an unbranched peduncle (sect. Ulei lacks the basal rosette in mature shoots). The inflorescence has spreading spathes during anthesis (except M. Juruanum which has the spathes rolled up around the inflorescence-rachis throughout flowering). The one-flowered cymules have short axes. The corolla tube is long and narrow, the lobes are oblong with cucullate apex and there is often one outer petaloid staminode (sometimes reduced).

These characters indicate an affinity with <u>Maranta</u> or <u>Marantochloa</u> rather than with <u>Ischnosiphon</u> which it was included in earlier. In both <u>Maranta</u> and <u>Marantochloa</u> there are species with comparatively short cymule axis and long corolla tubes.

The pollination syndrome is melittophily with long-tongued euglossine bees as legitimate pollinators. Autogamy probably occurs within the M. laxum-complex.

<u>Monotagma</u> can be subdivided into two large groups: one with the aril bilobed, and one with the aril entire. The Ecuadorian material is grouped as follows:

1. Species with a bilobed aril and $^+$ excentric leaf-blade apex.

<u>M. laxum</u> (P. & E.) Schum.
A highly polymorphic species with a wide distribution. It is an example of a so called "ochlospecies", which during the climatic changes with long periods of drought during Pleistocene survived as various allopatric populations in different forest refuges. The temporary reproductive isolation did not give rise to any genetic incompatibility. During the isolation some minor morphological changes did occur and when they redispersed the variation is no longer geographically well-correlated.
It can be recognized by its long green spathes and long, often pendent, inflorescences.

<u>M. Juruanum</u> Loes.
Readily recognized on its stout (spiciform) florescences with the spathes rolled up around the florescence rachis throughout flowering, and by the deep red flowers. It has a western Amazonian distribution.

2. Species with entire aril and $^+$ centric leaf-blade apex.

<u>M. secundum</u> (Peters.)Schum.
Recognized by the dense and ovoid headlike synflorescence

and the bright red spathes. It forms a species pair with M. duidae Steyerm.. They occur on different soils: M. secundum on latosols and M. duidae on white sand.

M. "rudanii", an unpublished taxon

Mature shoots lacks the basal leaf-rosette. It forms a species pair with M. ulei Schum. ex. Loes.. M. ulei has an eastern Amazonian distribution while M. "rudanii" is known only from the type collection from Zamora-Chinchipe.

3.4.2.4. L.B. Holm-Nielsen, AAU:

Distribution of the Alismatidae in the Neotropics

During the last few years I have been revising the families of the Alismatidae of the Neotropics and some regional floras in collaboration with Robert Haynes, University of Alabama.

Most of these plants are aquatics with generally wide distributions. They often appear to be pioneers or weeds, and offer no interesting data for the study of the history and phytogeography of an area. However, studies of distribution of the Alismatidae result in the following main patterns:

A. Temperate - High Andine
B. Lowland tropical
    1) Northern affinity.
      a) NW-South america b) central America c)carribean.
    2) Southern affinity
      a) SE Brazil b) Patagonia
    3) N-S bicentric
    4) Endemic
    5) Wide distribution

This is a general system very like that of many other taxa. The distribution maps pose a problem: Why are the Alismatidae relatively uncommon on the Amazonian basin? -or are aquatic plants scarce in the Amazon?

3.4.2.5. Flemming Skov, AAU:

Revision of the Palm Genus Hyospathe in Tropical South America

The genus Hyospathe comprises 18 species of small, understory palms according to C.F. Glassman in his Index of American Palms. The genus is widespread in wet forest in tropical South America ranging from Panama to Bolivia. There is a concentration of species on the eastern slopes of the Andes and in the western parts of Amazonas.

All species are confined to dense, primary forest from the lowland up to ca. 2000 m. They are seldom found outside virgin forest and depend on deep shade and high humidity to regenerate and survive.

No modern treatment of the genus exists and it seems to be in great need of a revision. Field work and preliminary examination of voucher specimens show that at least some of the species are very variable. Because many of the species, described in the genus are based on such characters as leaf morphology and size/shape of the inflorescence, a reduction in the number of names is expected.

Ecuador is comparatively rich in species.
The widespead <u>Hyospathe elegans</u> is found in Amazonian Ecuador and probably in the north-eastern coastal lowland as well. The pinnate-leaved <u>Hyospathe gracilis</u> is found in the south-eastern part of the Amazonas. On the eastern slopes of the Andes in the Puyo area two interesting species are found: <u>Hyospathe cf. schultzeae</u>, a very small palm with minute bifid leaves, and <u>Hyospathe macrorhachis</u>, a very outstanding member of the genus, which is set apart from the rest of the species in having pedicellate pistillate flowers and fruits. It is also very easy to recognize because of the very long rachis.

3.5.1. Jørgen Korning, AAU

New Results in the Anangu Rain Forest Project.

Four years after its initiation the SEF-project with continuing visits, totalling ten months of observations, the investigations have expanded to include soil analysis, ethnobotany, pollination biology and mycorrhiza studies. Still floristics and quantitative studies are the major subjects.

Since the 1984 workshop, Karsten Thomsen and I visited Anangu 3 times, continuing our study of rain forest ecology, emphasizing the subjects quantitative and qualitative species composition, and soil.

Our study site is a one hectare quadrat in which trees with DBH 10 cm or more were included. We measured DBH, height, and collected vouchers from trees which could not be identified in situ. Different kinds of collecting equipment has been applied. First the "tree-bicycle" (Swiss tree-grippers): Although it has the advantage of not hurting the trees, it is heavy to carry and use, and it is difficult to use on branched trees and on irregular wet or fluted trunks. Semicircular spurs are relatively light and easy to handle, although two sizes had to be taken along, in order to cover the variation of DBH. Unfortunately they leave some marks on the trunk. This may weaken the tree. We have not used ropes successfully. Our favorites are the spurs and - of course the monkey way.

To reach the branches we used a treepruner on a 6.5 m long cane of Gynerium sagittatum growing on the river bank. No other material exceeds its efficiency. To get good material from a tree it is important to check for flowers and fruits on the entire tree. Often we discovered flowers by luck on some few branches. Because most of the trees are sterile in the major part of their lifetime it is important that they can be identified on sterile material. Therefore it is of importance to be careful getting the most important sterile characters on the voucher. These could be light/shade exposed leaves, whole compound leaves, leaf position, stipules, shoots etc. and notes about bark slash and exudates. To get the best voucher one must dry the specimens immediately after collection. That means not using any sort of temporary conservation although this usually is difficult to avoid.

A comparison of the ten highest Family Importance Values from the three sample areas, floodplain, terra firma transect and terra firme quadrat shows a surprising difference between the two terra firme samples. While the floodplain and terra firme transects are more alike.

These differences show a forest of great variability and some limitations of the two sampling types. - There is much work to be done in the area yet.

3.5.2. Simon Lægaard, AAU:

Páramo Vegetation in Ecuador.

Grass páramos have generally been considered the natural vegetation of the altitudinal zone from app. 3200-3500 m to 4100-4300 m in the aequatorial Andes.

The occurrence to about 4100-4300 m of small stands of Polylepis forest has been explained by some very special condititons. According to one of these, Polylepis should only occur at these altitudes on boulder-screes, though which the cold air in the soil could be drained downwards. I have numerous observations, that also the highest stands of Polylepis grow in "ordinary" compact soil and that there are no significant differences in soil structure or in topography between these high stands of Polylepis and their neighbouring areas of grass páramos. Therefore I consider that the traditional explanation has to be abandoned.

Most paramos are regularly burned with a lapse of 1-2 years. It has ben found through many observations that Polylepis is very susceptible to fire. Every time a páramo fire reaches a forest border, a zone of app. 3-5(-10) m of Polylepis trees are killed by the fire.

In the ordinary grassparamo it has been found that all plant species occourring are pyrofytes, that is, they are adapted, in one way on another, to survive a fire.

A number of survival mechanisms have been observed and studied. These can be arranged in the following groups:

1. Survival of aerial stem and shoot-apex. Ex.: Puya, Esepletia, Blechnum, Valeriana plantaginea.

2. Survival by buds sprouting from stems in or near soil surface. A very varied group that may be divided in serveral subgroups. Ex.: Calamagrostis, Festuca, Paspalum bonplandianum, Gunnera magellanica, Ranunculus, Hypericum, Valeriana microphylla.

3. Survival by subterranean bulbs or rhizomes. Ex.: Orchids, Bomarea, Carex, Halenia.

4. Plants dying, but survival of species by seeds. Ex.: Lupinus, Bartsia.

My conclusions from these studies are, that most probably the whole area of the grass páramo zone has been covered by a forest that , at least in the upper part, was dominated by Polylepis.

3.5.3. Karsten Thomsen, AAU:

Preliminary results of floristic studies on terra-firme trees in Anango.

Serveral study periods totalling more than 20 weeks of observation and collection on the terra-firme SEF-line have added considerably to the number of vouchered trees. This has proved critical for the species picture, as half of the species are represented by only one individual. Moreover the follow-ups have brought up the percentage of species collected fertile from the initial less than 10% to now more that 50%.

As expected, the recent approx. doubling of study area have augmented the species number (some 28%) and the representation of rare species (less than 1 ind./ha.) Interestingly, the new plot also seems to have a remarkably different composition of dominating species, here the two most common SEF-line species being represented by only 2 individuals.

Futhermore, compared to the line transect our plot has significantly fewer species in common with the nearby floodplain forest, and seems to be homogenously dominated by more strict terra-firme species, whereas the terra-firme line seems more like a mosaic throughout its 4 km extension with certain species more common in the moist "quebradas". This may account for the higher species diversity of the line and raises some questions: Is a long transect or a square most representative for the tree vegetation ? how patchy is the species distribution?

Identification of the Anango trees has been greatly facilitated by the accumulating matching material - both vouchers, of which most are sterile, and general collections - and by phenological observations, which we have carried out 5 times on the terra-firme line with binoculars. A pair of good binoculars can be strongly recommended as it allowed us to recognize 56% of the trees on the hectare-plot without collection. Identification was further improved by a very high degree of fertility during our sampling period which exceeded anything we have seen on the SEF-line, undoubtedly due to the favorable light conditions at the plot, which is situated on a rather high plateau.

In spite of the various Anango study sites, which counts more than 400 tree species on less than 3.2 ha., the species list is probably still far from complete: more than 60% of the species are represented by only one individual, and so are even some interesting families: Hippocastanaceae, Magnoliaceae, Piperaceae, Proteaceae, Sabiaceae, Solanaceae, Staphyleaceae, Styracaceae, Theaceae and Urticaceae.

3.6.1. John Brandbyge, AAU

Reforestation by means of local tree species in rural disticts of Ecuador.

On the interandean plateau man has been interacting with the vegetation for up to 10.000 years. In Ecuador the Andean slopes were formerly covered by forests up to about 4.000 m, but today only small areas covered with high-andean forest remain. The deforestation continues with severe consequences for the inhabitants in rural areas. Destruction of the natural vegetation cover leads to fuel-wood shortage and soil erosion.

Fortytwo percent of rural households use fuel-wood as the only source of energy for cooking and 75% use fuel-wood as the only source for heating. Especially in the highest parts of the Andes the task of providing enough fuel-wood is hard. A family that relies solely on collected fuel-wood for energy uses 2 whole days per week to meet their demands.

Many parts of high-andean Ecuador are severely affected by soil erosion, and in certain dry areas a rapidly ongoing desertification can be observed. In a recent study it was calculated that on abandoned fields deprived of vegetationals cover the amount of top-soil lost due to soil erosion was 82.7 t/ha/y.

The answer to these problems of course is reforestation. Figures from 1983 tell that 4.500 ha or about 3% of the estimated 150.000 ha that were felled, was replaced by plantation of mostly exotic species. Planting of exotic species such as Pinus radiata and Eucalyptus globulus in Ecuador have problems of adaptation and have impoverished the local flora and soil.

Our project was started to find alternatives to the traditionally planted exotic species, among the vast number of local tree species, to test them and to introduce them into reforestation programmes in high-andean Ecuador. Buddleja incana, Vallea stipularis, Polylepis spp. Gynoxys spp., and Oreopanax spp. were set up, and small scale pilot plantations were established with promising results. The project was carried out in collaboration with the local indigenous population.

3.6.2. Lars Peter Kvist, AAU:

Aspects of Ecuadorean lowland Indian ethnobotany.

The ethnobotany of the Ecuadorean indigenous groups, located on the coastal plain and those of the northern Amazonian lowland, are reasonably well-known. Students and staff members of the Aarhus group have undertaken intensive ethnobotanical collecting with the Colorado, Cayapa, Coaiquer, Siona-Secoya and Quichua (Yumbo) Indians. Other workers have studied Cofan, Waorani and Siona-Secoya Indians.

It is interesting to compare and summarize plant uses of different indigenous groups. For instance medicinal plants are more likely to have physiological effects, when used for the same purpose among different groups with little contact. Here the plants used for birth-control and arrow-poisons are briefly discussed.

Birth-control. Especially two plants used for contraception and/or sterilization are intersting. These are:

1) Persea americana, avocado-seeds are used by the Colorado, Siona-Secoya, Quichua (Yumbo) and Tukuna Indians (from the border area af Brazil, Peru and Colombia). Crushed seeds may be eaten (Quichua) or a decoction may be drunk (Colorado, Tukuna).

2) Brownea spp., several species, e.g. B. grandiceps are know in Amazonian Ecuador as "cruz caspi". Quichua (Yumbo) and Siona-Secoya Indians mainly use decoctions of the heartwood bud also leaves and flovers. Quichua and Siona-Secoya Indians sometimes mix Persea, Brownea and other plants, e.g. leaves of Rudgea.

The present material suggests that the traditionally war-like Waorani and Shuar tribes have fewer birth-control plants, than the more peaceful groups Quichus, Siona-Secoya and Colorado.

Arrow poison. We may distinguish curare and non-curare poisons. The first ones are based on Menispermaceae and/or Strychnos.

Amazonian Ecuador. The primary ingredients are different Menispermaceae, mainly Curarea tecunarum and Chondrodendron tomentosum. Species of Strychnos are usually, but not always added, and many other plants are occasionally admixtured, e.g. different Annonaceae. Some may improve the poisonous effects, others the texture, durabillity ect.

Pacific Ecuador. Only non-curare poisons are known. All three tribes traditionally use Naucleopsis spp. The latex was applied to the darts without previous preparation. The Chocó Indians of Pacific Colombia also use Naucleopsis. Furthermore they use poisonous frogs, a practize unknown in Ecuador.

3.6.3. Bo Boysen Larsen, AAU:

Prospects of Quinua cultivation in the Andes.

Cultivation of native plant species has a very long history in the Andes, The plants utilized can be largely divided into two subgroups. The tuberous plant, including the potato (<u>Solanum tuberosum</u>), the oca (<u>Xalis tuberosa</u>) and the melloco (<u>Ullucus tuberosus</u>), provide mainly starch. The plants providing edible fruits or grains, on the other hand, are important sources of proteins and vitamins. These include the amaranths (<u>Amaranthus</u> spp.), the chocho (<u>Lupinus mutabilis</u>) and the quinua (<u>Chenopodium quinoa</u>).

The cultivation of these native crops has declined since the Spanish conquest and the subsequent introduction of cereals like wheat and barley. However, the native grain crops are undoubtedly well suited to combat malnutrition in the Andean countries.

The grains of <u>Chenopodium quinoa</u> have a fairly high protein content (ca. 14%) and the protein quality equals that of milk. The proportion of lysine, which is the limiting amino acid in wheat and rice, is very high and the grain is thus and ideal substitute in wheat flour.

Saponins are abundant in the pericarp of quinua grains and must be removed before consumption because of their bitter taste and hemolytic action. The saponins serve to protect the plants from birds.

Due to artificial selection by indigenous people, an array of different quinua strains exists. Plants with large grains have been selected. A great variation concerning saponin content, inflorescence size, drought tolerance and frost resistance is found. Germplasm is now collected in all countries along the Andes, and breeding experiments are carried out.

A project on quinua cultivation in Denmark is planned.

4.1 Lauritz B. Holm-Nielsen, AAU:

The Aarhus University Ecuador Project (AAU-Ecuador project) 1984-86

Systematic expeditions

In the period 1984-86 the Aarhus group has undertaken a number of expeditions to Ecuador most of which with a narrow scope directly related to an ongoing revision of a certain plant group. Ferns (Illum-Borchsenius, 1985), Cactaceae (Madsen, 1984,85), Passifloraceae (Jørgensen, 1984), Fungi (Læssøe 1985), Valerianaceae (Eriksen, 1985). A general survey of new areas was carried out in 1985 by Øllgaard and others.

Studies of Ecuadorean palms:

A new 5 year project of the systematics, biology and use of Ecuadorean palms, was launched in 1985 by Balslev et al. The first field campaign was carried out by Balslev, Barfod and Skov 1985; the second is ongoing (Balslev, & Borchsenius 1986).

Vegetation studies

The studies at the Anangu site continues (Øllgaard, Thomsen, & Korning 1985, Thomsen & Korning 1986). A new project on the structure and systematics of Ecuadorean mountain forests has just been started in collaboration with PUCE (Jørgensen,1986).

The Paramo use project has been concluded - report is under preparation (Lægaard).

Ethnobotanical field work (AAU-SEE project)
Field work has been continued in connection with other projects especially in the northwestern and the eastern lowlands.

Collaboration with the P. Universidad Catolica (PUCE), Quito

The close collaboration with PUCE is still the cornerstone in our Ecuadorean projects. Two members of the Aarhus group have been residents in Quito in the period 1984-86 (Simon Lægard), 1985-86 (Bo Boysen Larsen). From March 1986 Peter Møller Jørgensen has started his work as a staff member at PUCE.

Collaboration with the Charles Darwin Research Station (CDRS), Galapagos

In the mid seventies Stig Jeppesen was resident botanist at the CDRS. This relation was reinstated in July 1986 when Jonas Lawesson began his work in the Galapagos Islands.

Collaboration with Instituto Nacional de Energia (INE) and Centro Ecuatoriano de Servicios Agricolas (CESA)

At the end of 1985 the firewood project was concluded and reported on (Brandbyge & Holm-Nielsen 1985). It is intended to follow this project up together with INE/CESA and by receiving an Ecuadorean biologist for further training in Aarhus.

Research programs and related work
1. Systematic-taxonomic studies:

   a. Flora of Ecuador
      Lobeliaceae (S. Jeppesen) published
      Juncaceae (H. Balslev) published
      Alismatidae (L.B. Holm-Nielsen & R.R. Haynes) in press
      Anacardiaceae (A. Barfod) in press
      Passifloraceae (L.B. Holm-Nielsen, P.M. Jørgensen & J.E. Lawesson) submitted

Lycopodiaceae (B. Øllgaard)

Cactaceae (J.E. Madsen)

Valerianaceae (B. Eriksen & B. Boysen Larsen)

Arecaceae (H. Balslev)

Poaceae (S. Lægaard)

b. Revisions:

Triplaris (J. Brandbyge) in press

Blechnum (D. Nielsen & T. Illum)

Xylaria (T. Læssøe)

Hyospathe (F. Skov)

Phytelephas (A. Barfod)

2. Floristic-Phytogeographic studies
   a. Phytosociology:

Studies of Ecuadorean Forests. SEF-project. Anangu (Balslev, Øllgaard, Holm-Nielsen, Korning, Thomsen)

Paramo studies: Collaborator PUCE. (Balslev, Lægaard, Holm-Nielsen).

Mountain forests: Collaborator PUCE. (Øllgaard, Møller-Jørgensen).

Dry coastal areas: Collaborator CDRS (Lawesson, Holm-Nielsen

3. Ethnobotanic studies
   a. Firewood etc.: (Brandbyge, Holm-Nielsen)
      Collaborators INE, CESA, DANIDA, INTERCOOPERATION

b. Extensive use of paramos: Collaborators PUCE, DANIDA (Lægaard)
   Collaborators PUCE, DANIDA

c. Ethnobotany of Indigenous groups:
   Collaborator Banco Central.
   Colorados (Kvist, Holm-Nielsen).
   Gayapas (Kvist, Barfod, Holm-Nielsen).
   Coaiqueres (Barfod, Kvist, Nissen).
   Secoyas (Brandbyge, Balslev, Holm-Nielsen).
   Sionas (Brandbyge, Azanza, Barfod).
   Yumbos (Alarcon, Lawesson, Jørgensen).

Aucas (Jaramillo, Coello).

4. Educational program
   a. Residents in Quito:
      1979-81   L.B. Holm-Nielsen
      1981-84   H. Balslev
      1984-86   S. Lægaard
      1985-86   B. Boysen Larsen
      1986-   P. Møller Jørgensen

   b. Fellowships in Aarhus:
      1981-83   J. Jaramillo
      1981-83   F. Coello
      1984-85   B. León
      1985   P. Mena

5. Nature conservation

Advising of and collaboration with Ministerio de Agricultura
y Ganaderia, Fundación Natura, Museo Ecuatoriano de Ciencias
Naturales, Banco Central, PREDESUR, INCRAE, INGALA, Oficina
Fronterizo.

6. Publications 1984-1986.

Balslev, H. & Barfod, A. 1986:
  Ecuadorean Palms - an overview. Opera Bot. (in press).
  35 pp. ms.
Balslev, H. & Henderson, A. 1986: A new Amandra (Arecaceae)
  from Ecuador. Syst. Bot. (submitted).
Balslev, H., Holm-Nielsen, L.B. & Øllgaard, B., 1985:
  Paramoen - Andesbjergenes alpine vegetation. - Naturens
  Verden 1985: 363-376.
Balslev, H., Luteyn, J.L., Øllgaard, B. & Holm-Nielsen,
  L.B., 1986: Composition and structure of Adjacent
  Unflooded and Floodplain Forests in Amazonian Ecuador -
  Opera Bot. (in press). 60 pp ms.
Balslev, H. & Lægaard, S., 1986: Distichia acicularis sp.
  nov. - a new cushion-forming Juncaceae from Ecuador. Nord.
  J. Bot. 6: 151-156.

Barfod, A., 1986: Anacardiaceae.- In: Harling & Andersson: Flora of Ecuador (in press).

Barfod, A. & Holm-Nielsen, L.B. 1986: Two new Anacardiaceae from Ecuador.- Nord. J. Bot. 6: 423-426.

Barfod, A., Kvist L.P. & Nissen, D. 1985: Notes on shamanism of the Cayapas (Chachis) of NW Ecuador - Misc. Antropol. Ec. 5 (in press).

Brandbyge, J. 1984: Three new species of the genus Triplaris (Polygonaceae). - Nord. J. Bot. 4: 761-764.

Brandbyge, J., 1984: Especies Forestales Nativas En Los Andes Ecuatorianos. 1-50. CESA, Quito.

Brandbyge, J. 1985: El genero Triplaris (Polygonaceae) en el Ecuador. - Publ. Museo Ecuat. Cienc. Nat. Ser. Revista 6 (4): 9-19.

Brandbyge, J., 1986: A taxomomic revision of the genus Triplaris (Polygonaceae). - Nord. J. Bot. 6: 545-570.

Brandbyge, J. & Holm-Nielsen, L.B., 1986: Reforestation of the High Andes with local Species. - Reports Bot. Inst. Univ. Aarhus 13:1-114.

Brandbyge, J. & Øllgaard, B., 1984: Inflorescense structure and generic delimitation of Triplaris and Ruprechtia (Polygonaceae) - Nord. J. Bot. 4: 765-69.

Bravo Velasquez, E. & Balslev, H., 1985: Dinamica y adaptaciones de las plantas vasculares de los cienagas tropicales del Ecuador. - Reports Bot. Inst. Univ. Aarhus 11:1-50.

Haynes, R.R. & Holm-Nielsen, L:B.: A generic treatment of Alismatidae in the Neotropics, with special reference to Brazil. - Acta Amazonica suppl., 96 pp. (in press).

Holm-Nielsen, L.B. & Barfod, A., 1984: Las Investigaciones etnobotanicas entre Los Cayapas y los Coaiqueres. - Misc. Antropol. Ec. 4: 107-128.

Holm-Nielsen, L.B. & Brandbyge, J. 1986: Guds træ vender tilbage. - Udvikling, Danmark og U-landene 1986: 31-33.

Holm-Nielsen, L.B. & R.R. Haynes 1984: Two new Alismatidae from Ecuador and Peru. - Brittonia 37:17-21.

Holm-Nielsen,L.B. & Haynes, R.R. 1985: The identity of Limnocharis mattogrossensis O. Ktze. and its allies. - Phytologia 57:421-425.

Holm-Nielsen, L.B. & Haynes, R.R.: Alismataceae,

Limnocharitaceae, Hydrocharitaceae, Juncaginaceae, Potamogetonaceae, Zannichelliaceae, Najadaceae. - In: Harling &
Andersson: Flora of Ecuador,. (122 pp. ms. submitted).

Holm-Nielsen, L.B. & Jørgensen, P.M., 1986: Passiflora
tryphostemmatoides and its allies. - Phytologia 60:
119-124.

Holm-Nielsen, L.B., Jørgensen, P.M. & Lawesson, J., 1986:
Notes on Central Andean Passifloraceae. V: New species of
Passiflora subgen. Plectostemma and Tacsonia Nord. J. Bot.
(submitted.)

Holm-Nielsen, L.B. & Lawesson, J.E. 1986: New species of
Passiflora subgenus Passiflora. - Ann. Mo. Bot. Gard. (16
pp ms. in press).

Holm-Nielsen, L.B., Jørgensen, P.M. & Lawesson, J. 1986:
Passifloraceae. - In: Harling & Andersson: Flora of
Ecuador. (275 pp. ms. in press).

Holm-Nielsen, L.B., Øllgaard, B. & Molau, U. - (eds.) -
1984: Scandinavian Botanical Research in Ecuador.- Reports
Bot. Inst. Univ. Aarhus 9:1-83.

Jørgensen, P.M., Lawesson, J.E. & Holm-Nielsen, L.B., 1986:
A guide to collecting Passion flowers.-Ann. Mo. Bot. Gard.
7: 1172-1174.

Kvist, L.P. & Holm-Nielsen, L.B., 1986: Ethnobotanical
aspects of lowland Ecuador.- Opera Bot. (in press). 57 pp.
ms.

Larsen, B.B: 1986: A taxonomic revision of Phyllactis and
Valeriana Sect. Bracteata.- Nord. J. Bot. 6: 427-446.

Mena, P. & Balslev, H. 1986: Comparacion entre la Vegetation
de los Páramos y el Cinturón Afroalpino.- Reports Bot.
Inst. Univ. Aarhus 12:1-54.

Øllgaard, B., 1985: Observations on the ecology of
hybridization in the clubmosses (Lycopodiaceae). - Proc.
Roy. Soc. Edinburgh 86B: 245-251.

Øllgaard, B., 1985: Lycopodiaceae, pp. 148-159. In: A.R.
Smith: Pteridophytes in Venezuela, an annotated list. -
University of California, Berkeley.

Øllgaard, B. 1986: A revised classification of the
Lycopodiaceae sens. lat. - Opera Bot. (in press). 104
pp.ms.

AAU FIELD ACTIVITIES IN ECUADOR 1984-1986

AAU-exp. No.	AAU-residents	Other residents	Project	Sites	Map	Participants	Collections
17			Systematic	Andes	11	Illum,Borchsenius	56510-56999
18			General,SEF	Anangu,Andes	12	Øllgaard	57000-58606
19			SEF	Anangu	12	Thomsen,Korning	58611-58743
20			Systematic	Andes	13	Eriksen	59001-59423
21			Systematic	Oriente	17	Læssøe	59501-59999
22			Systematic	Lowlands	14	Barfod,Skov	60000-60150
22					14	Balslev	60500-60760
21					17	Læssøe	60801-60804
23	4		Systematic		15	Madsen	61000-61200
			SEF,General	Andes		Jørgensen	61201-65200
			Systematic	Lowlands		Balslev	65201-
24		3	General	Andes	16	Boysen	
		4	General	Galapagos		Laweson	

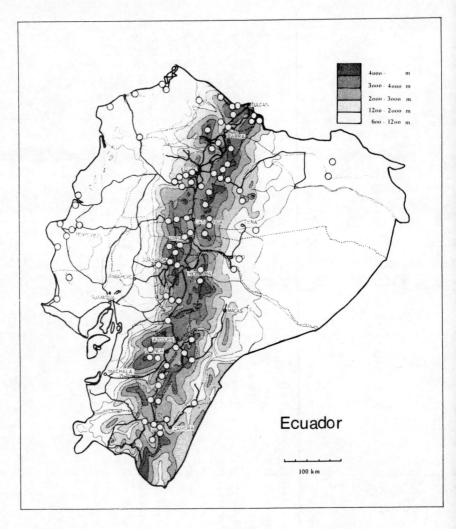

Collection localities of Simon Lægaard 1984 - 1985.
(51000 - 55999)

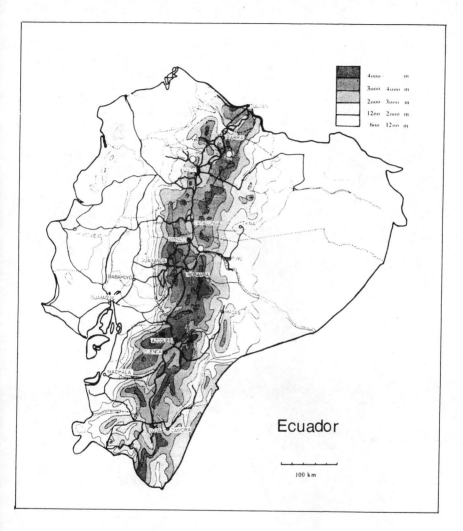

Collection localities of Illum, Borchsenius. 1985
(no. 5651o - 56999)

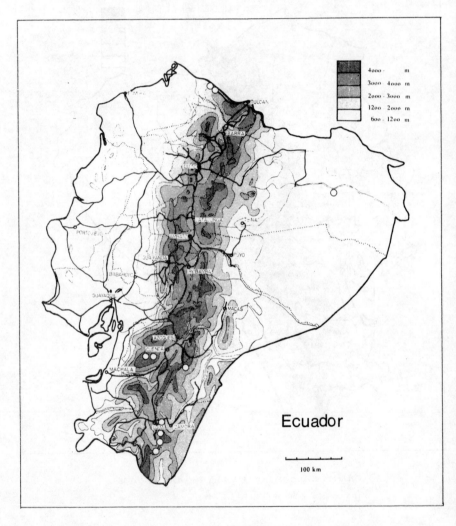

Collection localities of Benjamin Øllgaard, Jørgen Korning,
Karsten Thomsen, and Thea Illum 1985 (no. 57000 - 58743).

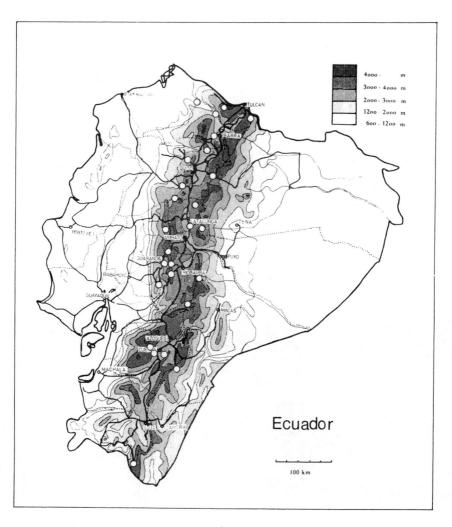

Collection localities of Bente Eriksen 1985
(no. 59001 - 59423)

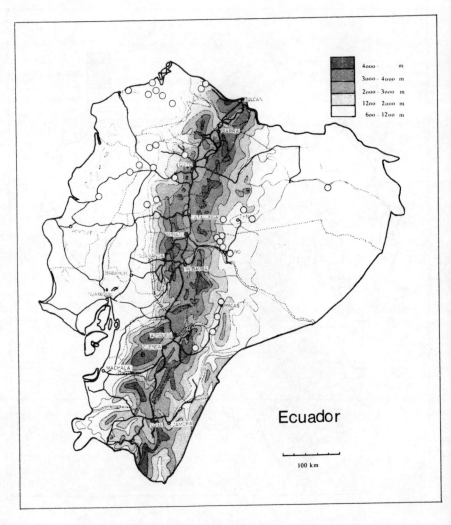

Collection localities of Henrik Balslev, Anders Barfod and
Flemming Skov 1985 (no. 60000 - 60760)

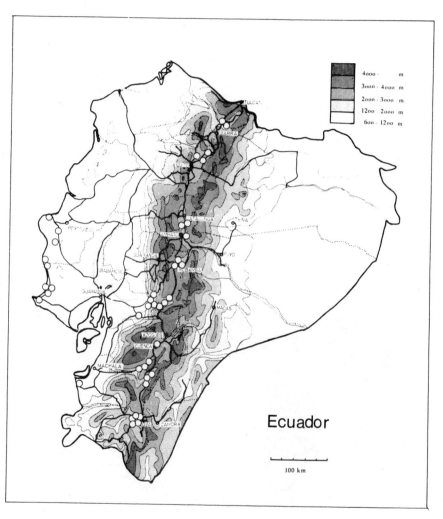

Collection localities of Jens Madsen 1985
(no. 61000 - 61200)

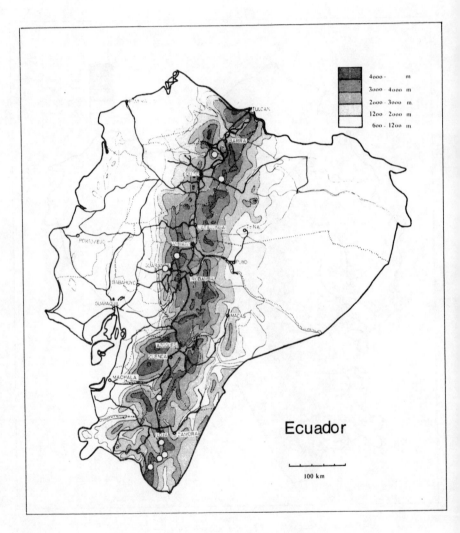

Collection localities of Bo Boysen Larsen 1985

- 73 -

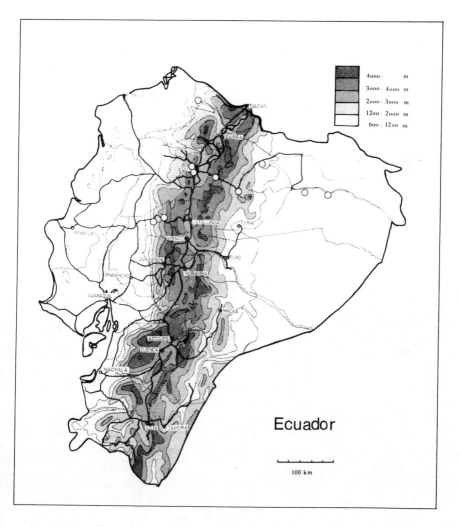

Collection localities of T. Læssoe 1985
(No. 59501-59999 og 60801-60804)

4.2. Ulf Molau, GB:

The Gothenburg University Neotropical Program 1984-86.

Flora of Ecuador.

The Flora of Ecuador has been continuously published under the editorship of prof. Gunnar Harling (see separate paper in the volume). After the unfortunate death of our good friend and colleague Baron Benkt Sparre in April 1986, Dr. Lennart Andersson, head curator of the Botanical Museum, has now entered the position as co-editor of the Flora, and B. Sc. Bertil Ståhl as assistant editor. THe publishing of the Flora of Ecuador is sponsored by the Swedish Natural Science Research Council.

Some of the critical groups of the Ecuadorian flora have proved to be suitable subjects for thesis works, and are now being revised by graduate students at our department. Those studies are enlarged to comprise entire groups in an often much wider area than Ecuador itself, and field work has been carried out in neighbouring countries as well. Current revisionary woirk carried out at the Department of Systematic Botany will be listed below.

Field work.

In the period 1984-86 the Neotropical Research Group at GB has carried out a number of expeditions to Ecuador and other Neotropical countries, many of which have been directly related to the revision of certain plant groups. A major expediton for the purpose of general collecting was undertaken during January-June 1985 by G. Harling and L. Andersson (with assistance of R. Eriksson, M. Hagberg and B. Ståhl), resulting in more than 5.000 numbers of Ecuadorian plant collections. Field work on several genera of the Scrophulariaceae was undertaken by U. Molau (assisted by L.Öhman) in Ecuador and Peru during three months in early 1985. In 1986, L. Andersson and M. Hagberg spent several months in Brazil to study certain groups of the Scitaminae. Other Neotropical areas visited during the period include

Hispaniola (M. Lindström and B. Ståhl, 1985) and Jamaica (U. Eliasson, 1985).

General lichen collecting as well as more specialized field work was carried out in January-March 1985 in Ecuador by L. Arvidsson, M. Lindström, and M. Lindqvist.

Current research project groups.

A number of project groups can be discerned within the Neotropical Research Group at GB, namely:

The Scitaminae Research Group (L. Andersson, M. Hagberg)

The Scrophulariaceae Research Group (U. Molau, F. Astholm)

The Ecuadorian Macrolichen Flora Group (L. Arvidsson, M. Lindqvist, M. Lindström).

Research program

1. Flora of Ecuador
    Cyclanthaceae (Gunnar Harling) published.
    Bixaceae, Cochlospermaceae, Elatinaceae (Ulf Molau) published.
    Scrophulariaceae: Calceolaria (Ulf Molau) published.
    Musaceae (Lennart Andersson) published.
    Amaranthaceae (Uno Eliasson) submitted.
    Alstroemeriaceae (Magnus Neuendorf).
    Araliaceae (Lennart Andersson).
    Centrospermae (excl. Nyctaginaceae) (Uno Eliasson).
    Compositae-Mutisieae (Gunnar Harling).
    Gleicheniaceae (Suzanne Roth).
    Marantaceae (Lennart Andersson).
    Nyctaginaceae (Jan-Eric Bohlin).
    Theophrastaceae (Bertil Ståhl).

2. Revisions

Calceolarieae (Scrophulariaceae; Flora Neotropica; Ulf
   Molau) submitted.

Marantaceae (Flora Neotropica; Lennart Andersson).

Erioderma (Pannariaceae; Lars Arvidsson).

Alonsoa (Scrophulariaceae; Fanny Astholm).

Colignonia (Nyctaginaceae; Jan-Eric Bohlin).

Neéa (Nyctaginaceae; Jan-Eric Bohlin).

Sphaeradenia (Cyclanthaceae; Roger Eriksson).

Monotagma (Marantaceae; Mats Hagberg).

Sticta (Lobariaceae; Mats Lindqvist).

Leptogium (Collemataceae; Marie Lindström).

Bartsia (Scrophulariaceae; Ulf Molau).

Bomarea (Alstroemeriaceae; Magnus Neuendorf).

Gleicheniaceae (Suzanne Roth).

Clavija (Theophrastaceae; B. Ståhl).

Publications 1984-86

Andersson, L., 1984. Sanblasia, a new genus of Marantaceae.
   Nord. J. Bot. 4: 21-23

Andersson, L., 1984. Notes on Ischnosiphon (Marantaceae).
   Nord. J. Bot. 4: 25-32

Andersson, L., The Chromosome number of Heliconia
   (Musaceae). Nord. J. Bot. 4: 191-193.

Andersson, L., 1984. (754) Proposal to reject the name Musa
   humilis Aubl. (Musaceae). Taxon 33: 524-525.

Andersson, L. 1985. Revision of Heliconia subgen. Stenochla-
   mys (Musaceae-Heliconioideae). Opera Bot. 82: 1-124.

Andersson, L., 1985. The identity of Ctenophrynium K. Schum.
   (Marantaceae). Nord. J. Bot. 5: 61-63.

Andersson, L., 1985. Musaceae. In G. Harling & B. Sparre
   (eds.), Flora of Ecuador 22: 1-84.

Andwrsson, L. & H. Kennedy, 1986. Four new species of
   Calathea (Marantaceae) from French Guiana. Nord. J. Bot.
   6: 000-000 (in press).

Andersson, L. Revision of Maranta subgen. Maranta
   (Marantaceae). Nord. J. Bot. (accepted for publ.).

Andersson, L. Marantaceae. In R. Dahlgren & P. Goldblatt (eds.), Families and Genera of Monocotyledons (submitted).

Andersson, L. & H. Kennedy. Marantaceae. In G. Harling & L. Andersson (eds.), Flora of Peru (in manuscript).

Eliasson, U., 1984. Native climax forests in the Galapagos Islands. In Perry (ed.), Galapagos. Key Environments Series, Vol. 1: 101-114. Pergamon Press, Oxford.

Eliasson, U., 1984. Chromosome number of Macraea laricifolia Hooker fil. (Compositae) and its bearing on the taxonomic affinity of the genus. Bot. J. Linn. Soc. 88: 253-256.

Eliasson, U., 1985. Identity and taxonomic affinity of some members of the Amaranthaceae from the Galapagos Islands. Bot. J. Linn. Soc. 91: 415-433.

Eliasson, U. Amaranthaceae. In G. Harling & L. Andersson (eds.), Flora of Ecuador (in press).

Eliasson, U. Morphological characters and taxonomic relations among the genera of Amaranthaceae in the New World and in the Hawaiian Islands. Bot. J. Linn. Soc. (submitted).

Harling, G. Cyclanthaceae. In R. Dahlgren & P. Goldblatt (eds.), Families and Genera of the Monocotyledons (submitted).

Holmgren, N. B. & U. Molau, 1984. Scrophulariaceae. In G. Harling & B. Sparre (eds.), Flora of Ecuador 21: 1-188.

Molau, U., 1984. New taxa and combinations in Calceolaria (Scrophulariaceae) from Peru and Bolivia. Nord. J. Bot. 4: 629-654.

Molau, U. & I. Sánchez Vega, 1986. Calceolaria (Scrophulariaceae): Las especies del Dpto. de Cajamarca, Peru. Boletin de Lima 8(43): 37-51.

Molau, U. Scrophulariaceae - tribe Calceolarieae. In G. Prance (ed.), Flora Neotropica (submitted).

Ståhl, B., 1986. Two new species of Clavija (Theophrastaceae) from NW South America. Nord. J. Bot. 6: 000-000 (in press).

Ståhl, B. Theophrastaceae. In R. Spichiger & G. Bocquet (eds.), Flora del Paraguay (in press).

ECUADOR

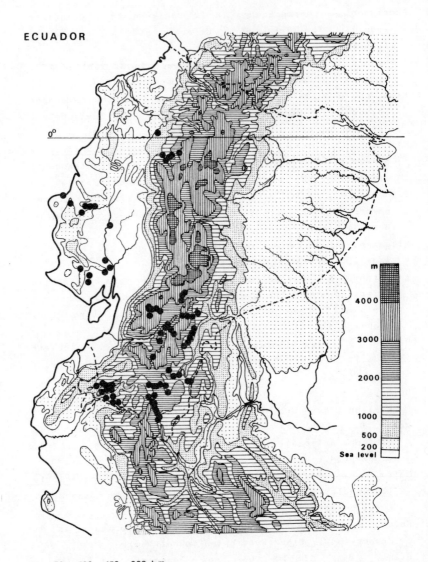

0   50   100   150   200 km

Collection localities of Gunnar Harling & Lennart Andersson, with the assistance of Roger Eriksson, Mats Hagberg & Bertil Ståhl, 1985

ECUADOR

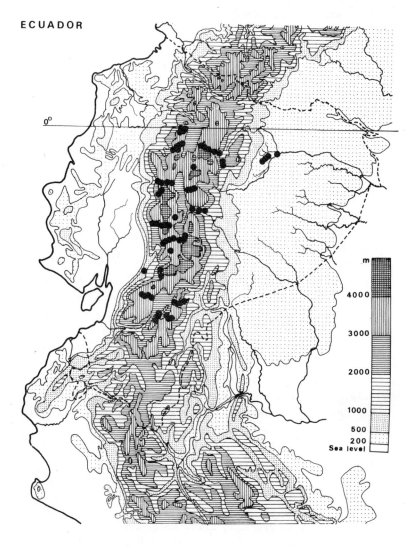

0 50 100 150 200 km

Collection localities of Ulf Molau & Lotte Öhman 1985

4.3. A cumulative map of the collection localities of Scandinavian Botanists in Ecuador.

After many years of field work in Ecuador, where collection sites were chosen because of recarch interests in a particular plant group or in a particular type of vegetation, it occurred to us that the compilation of a map showing all collection localities could help to indicate areas of future high priority. The following map shows where the approximately 120,000 collections by Scandinavian botanists were made. Who made which dots can be seen from other maps in this publications and from the maps in the proceedings of the first meeting (Reports Bot. Inst. Univ. Aarhus 9, 1984). Virtually all collections were made after 1939.

The rather intensive blackening of some areas generally indicates areas of easy access. These are generally the best known areas, but hardly any area in Ecuador can be claimed to be even fairly adequately known floristically. Careful collection is still desirable almost everywhere in Ecuador. Although a single dot may represent very few to more than a thousand collections, and thus is a somewhat unreliable expression of the actual amount of data from the site, - the map does indeed show where nothing has been done.

Judging from the present map we can discern five areas in particular need of botanical exploration:

1: The Sangay - Tiocajas area.
2: Cordillera de Cutucú - Cordillera del Condor to Zumba.
3: Lower Western slopes of almost the entire Cordillera Occidental.
4: Easten Slopes of the Cordillera Oriental.
5: Upland Manabí.

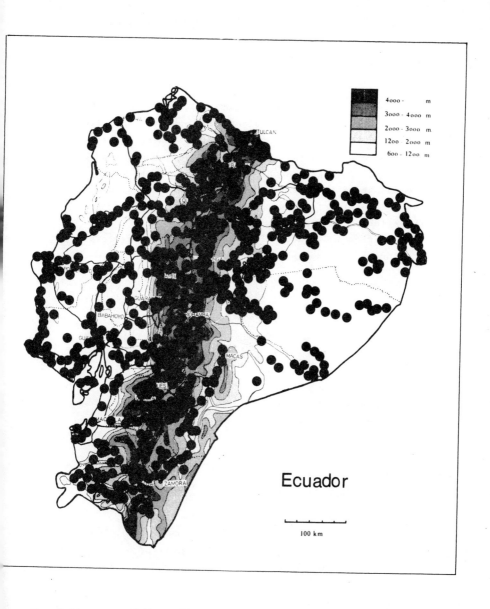

Cumulative map of the collection localities of Scandinavian
Botanists in Ecuador.

5. Lauritz B. Holm-Nielsen, AAU: Concluding Remarks:

What has happened and what is scheduled to happen in the
nearest future within the GB-AAU group.

In 1984 we presented 15 papers on the taxonomy of higher
plants and further 7 ethnobotany, ecology etc. The situation
in 1986 is 10 papers on taxonomy and 10 others. This is not
a sign of a decreasing effort put into taxonomic studies. We
have taken up larger taxonomic groups in recent years, and
have adopted a broader perspective in our work with more
students working in areas marginal to taxonomy, as
ethnobotany, ecology etc.
In the period 1984-86 the following families were completed
for Flora of Ecuador by members of our group.

Musaceae	45 sp
Anacardiaceae	18 sp
Scrophulariaceae	115 sp
Alismatidae (7 fam.)	45 sp
Passifloraceae	80 sp
11 families with	303 species.

In 1986 the following families are under treatment. The year
of expected completion is indicated in brackets.

Cactaceae	30	(1986)
Valerianaceae	40	(1986)
Theophrastaceae	11	
Amaranthaceae	100	
Lycopodiaceae	80	(1986)
Gesneriaceae	225	(1987)
Alstroemeriaceae	40	
Poaceae	450	new
Arecaceae	125	(1989)
Nyctaginaceae	45	
Mimosaceae	350	new
Asteraceae (Mutisieae)	50	
Centrospermae	50	
Polygonaceae	50	(1987)
Maranthaceae	100	
15 + families	ca. 1746 species	

Even with this increased commitment, with ecology projects
in most ecological zones of Ecuador, with ethnobotanical
projects among most of Ecuadorean lowland Indians we must
realize that we have an almost endless road of botanical
exploration and investigation in front of us.
In order to meet this challenge we must strengthen our
cooperation both within Scandinavia and with botanists
throughout the world. It is encouraging to participate in
this joint effort, and to realize the great promises of an
exceptionally large group of young botanists actively
participating in the work and receiving their training in
tropical botany. The colletive knowledge of Ecuadorean
plants and vegetation gathered by us is not matched
elsewhere.

6. Staff and students at the Botanical Institute

University of Aarhus, and the Department of Systematic Botany, University of Göteborg, participating in the Ecuador project, with their research intersts stated. Those marked with * participated in this meeting.

6.1. University of Aarhus.

Permanent staf:

*Lauritz B. Holm-Nielsen mag. scient, taxonomy of Alismatidae and Passifloraceae, Phytogeography, ethnobotany, applied botany.

*Benjamin Øllgaard, mag. scient., Taxonomy of Pteridophytes, especially Lycopodiaceae, vegetation studies.

Henrik Balslev, Ph.D., taxonomy of the Juncaceae and Palms, ethnobotany, vegetation studies.

*Simon Lægaard, cand. mag., páramo vegetation, taxonomy of Poaceae.

Ivan Nielsen, lic. scient, taxonomy of the Mimosaceae.

Temporary and associated staff.

*Anders Barfod, cand. scient., taxonomy of the Anacardiaceae and Palms, ethnobotany.

*John Brandbyge, cand. scient, taxonomy of the Polygonaceae, Reforestation.

Peter Møller Jørgensen, cand. scient., taxonomy of the Passifloraceae, montane forest.

*Lars Peter Kvist, cand. scient., taxonomy of the Gesneriaceae. Ethnobotany.

*Bo Boysen Larsen, cand. scient., taxonomy of the

Valerianaceae.

Jonas E. Lawesson, cand. scient., taxonomy of the
Passifloraceae. Vegetation of the Galapagos Islands.

Students

Thea Illum, cand. scient, taxonomy of Blechnum sect.
Lomariocycas.

Dorthe Nissen, cand.scient., taxonomy of Blechmum sect.
Lomaridium.

*Ulla Blicher-Mathiesen, economic botany of Bactris

*Bente Eriksen, taxonomy of the Valerianaceae.

Kirsten Gludsted, grass taxonomy.

*Isabelle Grignon, taxonomy of Sporobolus (Poaceae).

*Jørgen Korning, rain forest ecology.

*Thomas Læssøe, mycology, especially taxonomy of the
Clavariaceae.

*Jens E. Madsen, taxonomy of the Cactaceae.

*Flemming Skov, taxonomy of the palm genus Hyospathe.

*Karsten Thomsen, rain forest ecology.

6.2. University of Göteborg.

Permanent staff.

*Gunnar Harling, Professor, Dr., taxonomy of
Cyclanthaceae, Musaceae, and Compositae-Mutisieae;
phytogeography.

Uno Eliasson, Dr., taxonomy of Centrospermae, especially

Amaranthaceae; Myxomycetae.

*Lennart Andersson, Dr., Curator, taxonomy of Scitaminae, especially Marantaceae and Musaceae.

*Ulf Molau, Dr., Taxonomy and pollination biology of the Scrophulariaceae.

*Magnus Neuendorf, B. Sc., taxonomy of <u>Bomarea</u> (Alstroemeriaceae)

<u>Associated staff</u> (Göteborg)

*Lars Arvidsson, Dr. taxonomy of lichens especially macrolichens of Ecuador. At Nature Conservation Unit, City of Göteborg.

Students (Göteborg)

*Fanny Astholm, B. Sc., biosystematics of <u>Alonsoa</u> (Scrophulariaceae).

*Jan-Eric Bohlin, B. Sc., taxonomy of Nyctaginaceae.

*Roger Eriksson, B. Sc., taxonomy of <u>Sphaeradenia</u> (Cyclanthaceae).

*Sven Fransén, B. Sc., taxonomy of Bartramiaceae (Bryophyta).

*Mats Hagberg, B. Sc., taxonomy of <u>Monotagma</u> (Maranthaceae).

*Mats Lindquist, B. Sc., taxonomy of <u>Sticta</u> (Lichenes).

*Marie Lindström, B. Sc., taxonomy of <u>Leptogium</u> (Lichens).

*Suzanne Roth, B. Sc., taxonomy of Gleicheniaceae.
 *Bertil Ståhl, B., taxonomy of Theophrastaceae.